Max the Mouse
Fun with Numbers

Written & Illustrated by
A. S. Brightfield

Copyright © 2023 A.S. Brightfield

All rights reserved.

No part of this publication may be reproduced, distributed, or transmitted in any form or by any means, including photocopying, recording, or other electronic or mechanical methods, without the prior written permission of the publisher, except as permitted by U.S. copyright law.

The story, all names, characters, and incidents portrayed in this production are fictitious. No identification with actual persons (living or deceased), places, buildings, and products is intended or should be inferred.

Book Cover by A.S. Brightfield
Illustrations by A.S. Brightfield

ISBN: 9798396317543

For Halie, Caitlin, and Ashleigh, my treasures so true,
This book is dedicated to each one of you.
With hearts full of love and spirits so bright,
You fill every moment with pure delight.
May these pages spark your dreams and passions anew,
As you journey through life and all that you pursue.

In a cozy little house, Max loves to count, with glasses perched on his nose, he's always looking for the amount.

Max tiptoes through the
meadow, full of cheer,
he spots a single flower,
the number one, so clear!

Two shiny acorns make Max's whiskers twitch, he counts them with excitement, it's a perfect matching pitch.

2
two

Colorful balloons floating in the sky,
Max can't help but giggle,
as number three flies by.

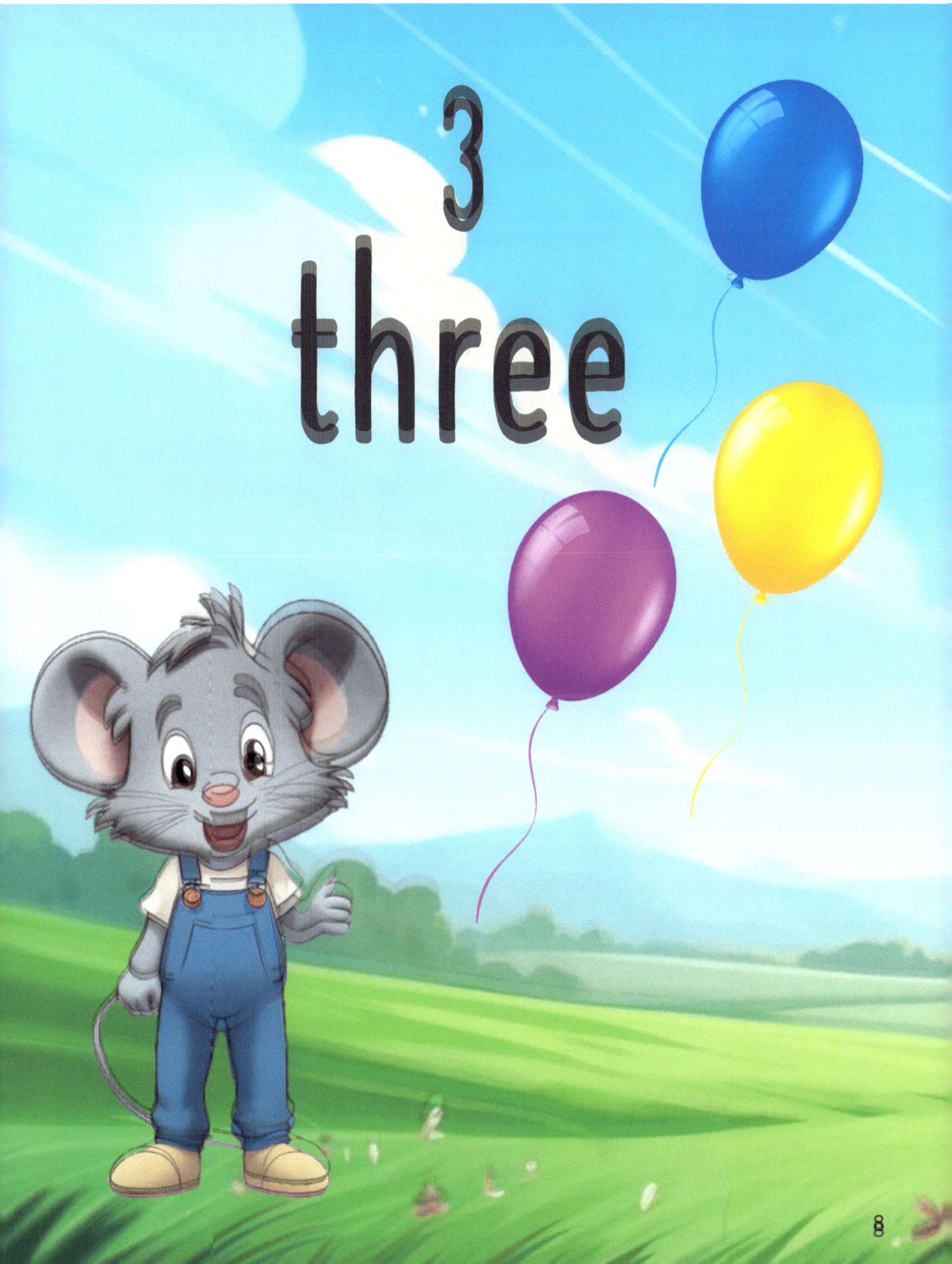

Butterflies flutter,
graceful in their flight,
four brings harmony,
a colorful sight.

Stars shining bright,
oh, what a splendid view,
counting up to five
is a joy Max always knew.

Six blocks stacked tall,
a tower of fun,
Max's imagination soars,
there's much to be done.

Juicy berries,
oh, so divine,
number seven brings
sweetness,
a taste that's truly fine.

Eight playful ladybugs
dance with delight,
Max joins in the fun,
it's a captivating sight.

Nine sparkling gemstones,
each one a work of art,
Max takes a moment to
admire them,
they touch his little heart.

9
nine

Ten soaring kites,
dancing in the breeze,
Max's heart soars with joy,
as they float by with ease.

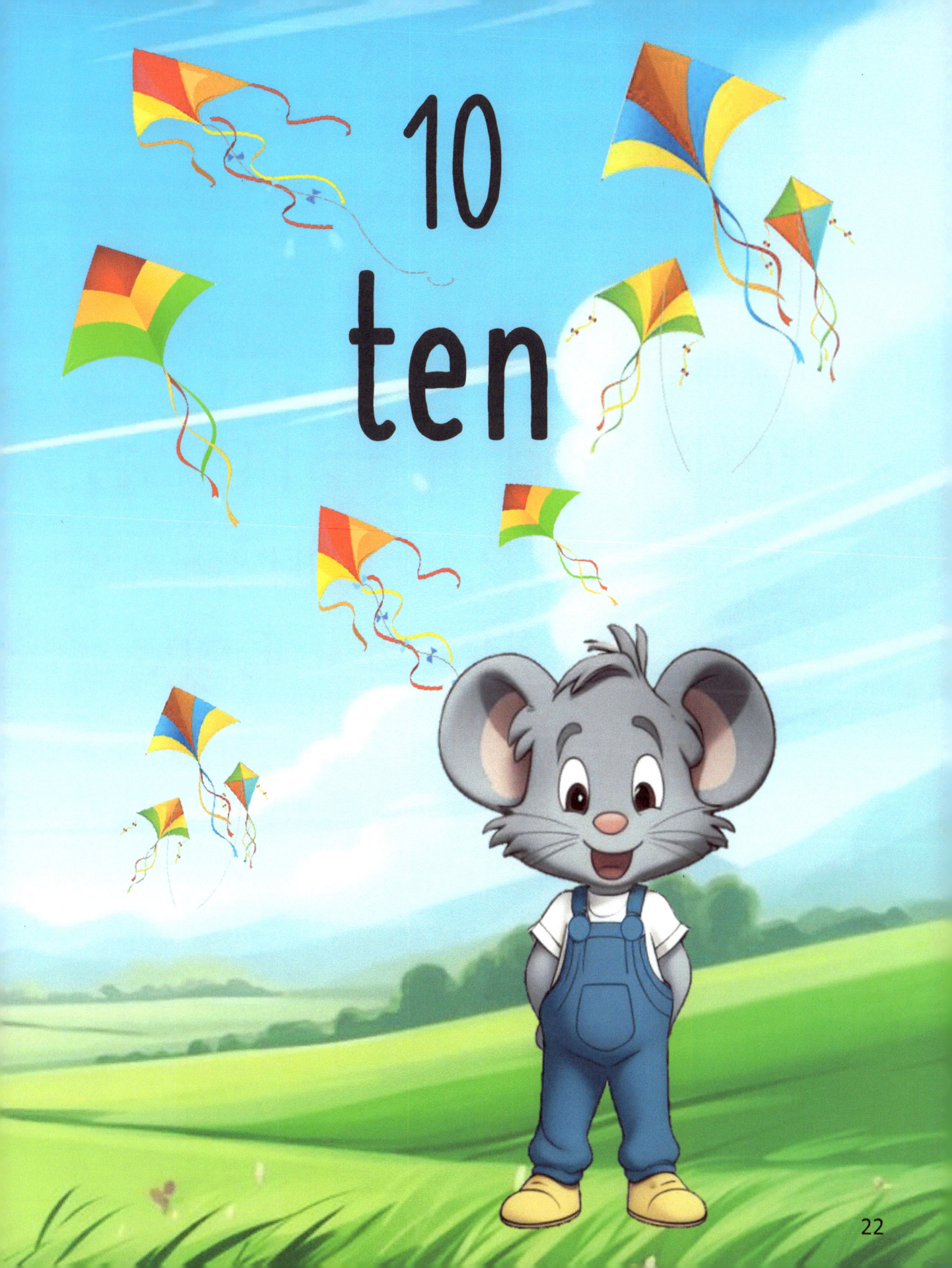

Reflecting in his house,
cozy and snug,
Max knows numbers
bring magic,
like a friendly hug.

Max invites all to join,
counting with glee,
exploring the world of numbers,
together we'll see!

Farewell for now,
but more adventures await,
with numbers as our guide,
we'll reach a delightful state.

Max waves goodbye,
surrounded by numerals
so bright,
let numbers lead the way,
filling our hearts with
delight!

www.ingramcontent.com/pod-product-compliance
Lightning Source LLC
Chambersburg PA
CBHW051940210526
45473CB00006B/2317